Flying Dragons

Ancient Reptiles That Ruled the Air

Written by David Eldridge
Illustrated by Norman Nodel

Troll Associates

Pronunciation Guide

Allosaurus	(Al-uh-SAWR-us)
Ctenochasma	(Ten-uh-KAZ-muh)
Dimorphodon	(Die-MORE-fuh-don)
Pteranodon	(Tuh-RAN-uh-don)
pterodactyl	(ter-uh-DAK-til)
pterosaur	(TER-uh-sawr)
Rhamphorhynchus	(Ram-fuh-RINK-us)
Tylosaurus	(Tie-luh-SAWR-us)

Printed in the United States of America. Troll Associates, Mahwah, N.J.
Library of Congress Catalog Card Number: 79-87965
ISBN 0-89375-241-X (0-89375-245-2 soft cover ed.)

Millions of years ago, reptiles ruled the world. They were everywhere. Gigantic dinosaurs roamed the land. Strange sea-going reptiles swam in the oceans. But strangest of all were the huge flying creatures that soared through the prehistoric skies. Like the dinosaurs and sea monsters, these "flying dragons" were reptiles.

One of the first of the reptiles to "fly" was a little animal that glided through the air. It lived over 200 million years ago—at the same time as the early dinosaurs. It could glide from the branch of one tree to another. It could also glide down to the ground, much as flying squirrels do today.

But, as time passed, other reptiles took to the air. They became better fliers. Their wings grew longer, and their muscles grew stronger. They began to flap their wings, and fly like modern birds. Some grew so large that they could truly be called "flying dragons."

The world of the "flying dragons" was a strange one. It was far different from the world we know today. In the age of reptiles, much of the earth was warm all year round. Huge ferns and palm-like trees grew everywhere.

The ancient seas were shallow, warm, and filled with fish. And the ancient skies swarmed with all kinds of flying reptiles — just as the skies today are filled with flocks of birds. Soaring and swooping, these reptiles spent their lives in a never-ending search for food.

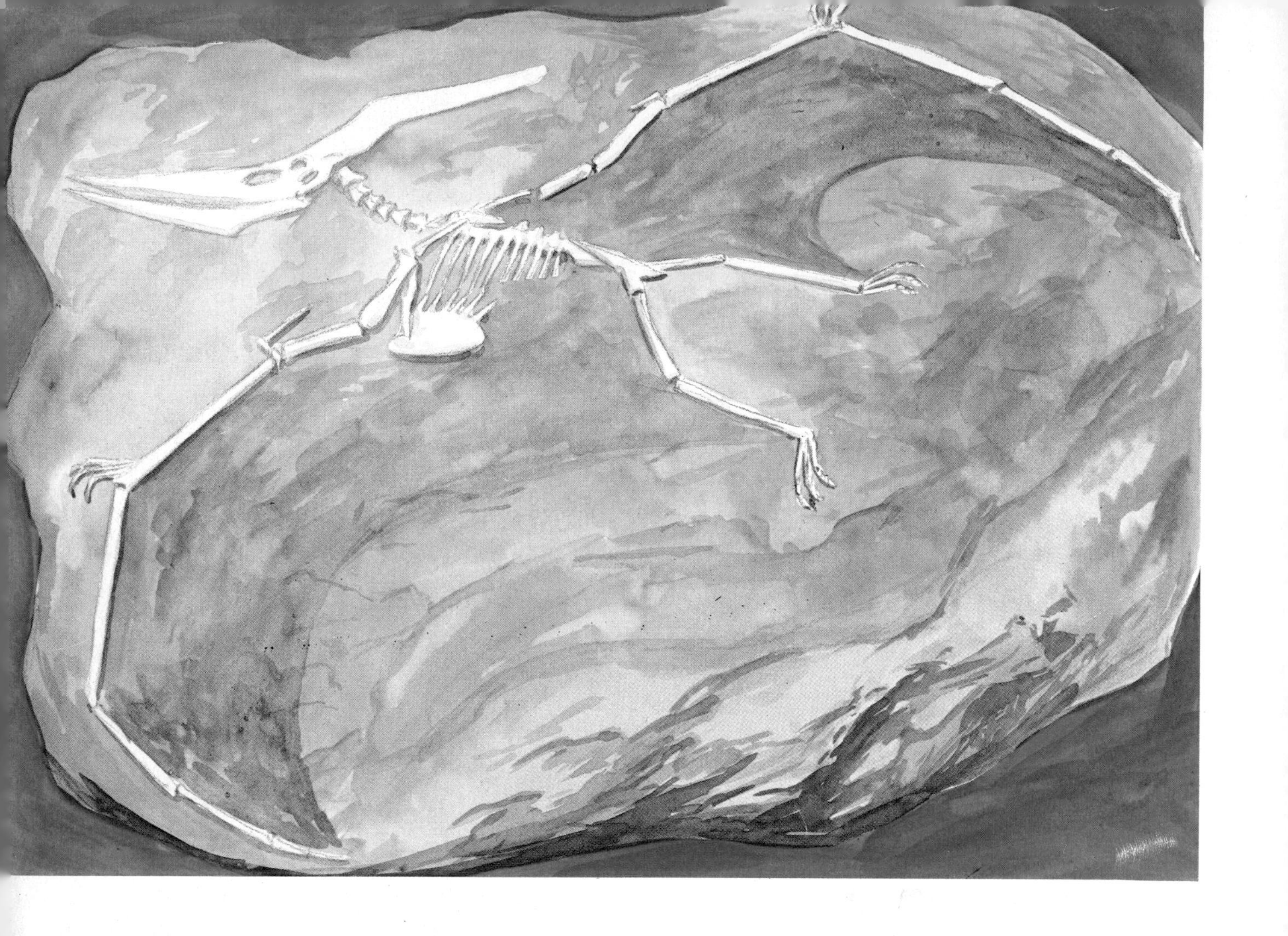

The reptiles that took to the air belong to one *order,* or group, of animals. Scientists call them pterosaurs, or "winged reptiles." There were many different kinds of pterosaurs, but they all had wings. Each wing was a web of leathery skin that was attached to the reptile's body. It stretched from the long "pinky" finger to the hind leg. The other fingers were small, with hook-like claws.

A typical pterosaur had a large head and long jaws full of sharp teeth. A pterosaur's teeth slanted forward, so it could drive them deep into its prey in mid-flight. Scientists think the early pterosaurs ate large, soft-bodied insects.

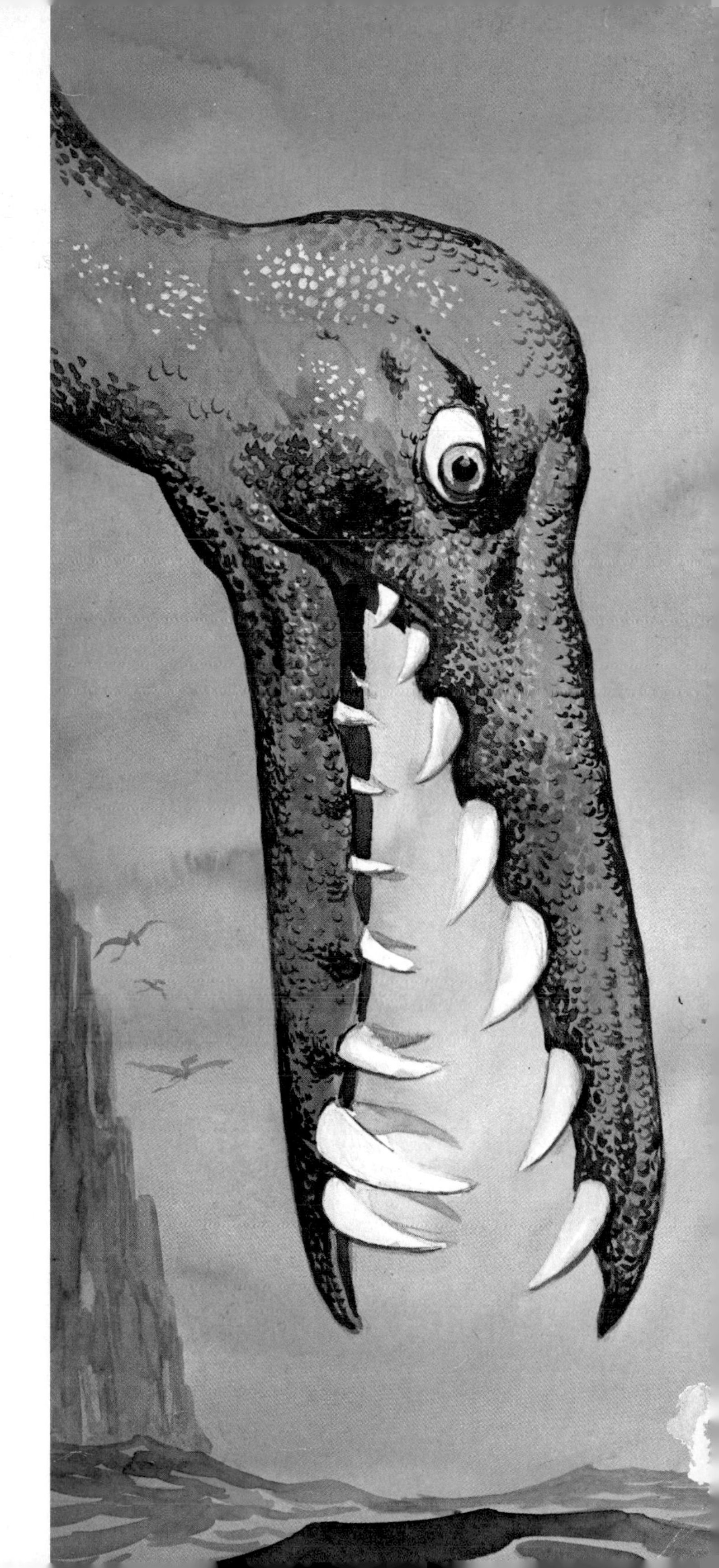

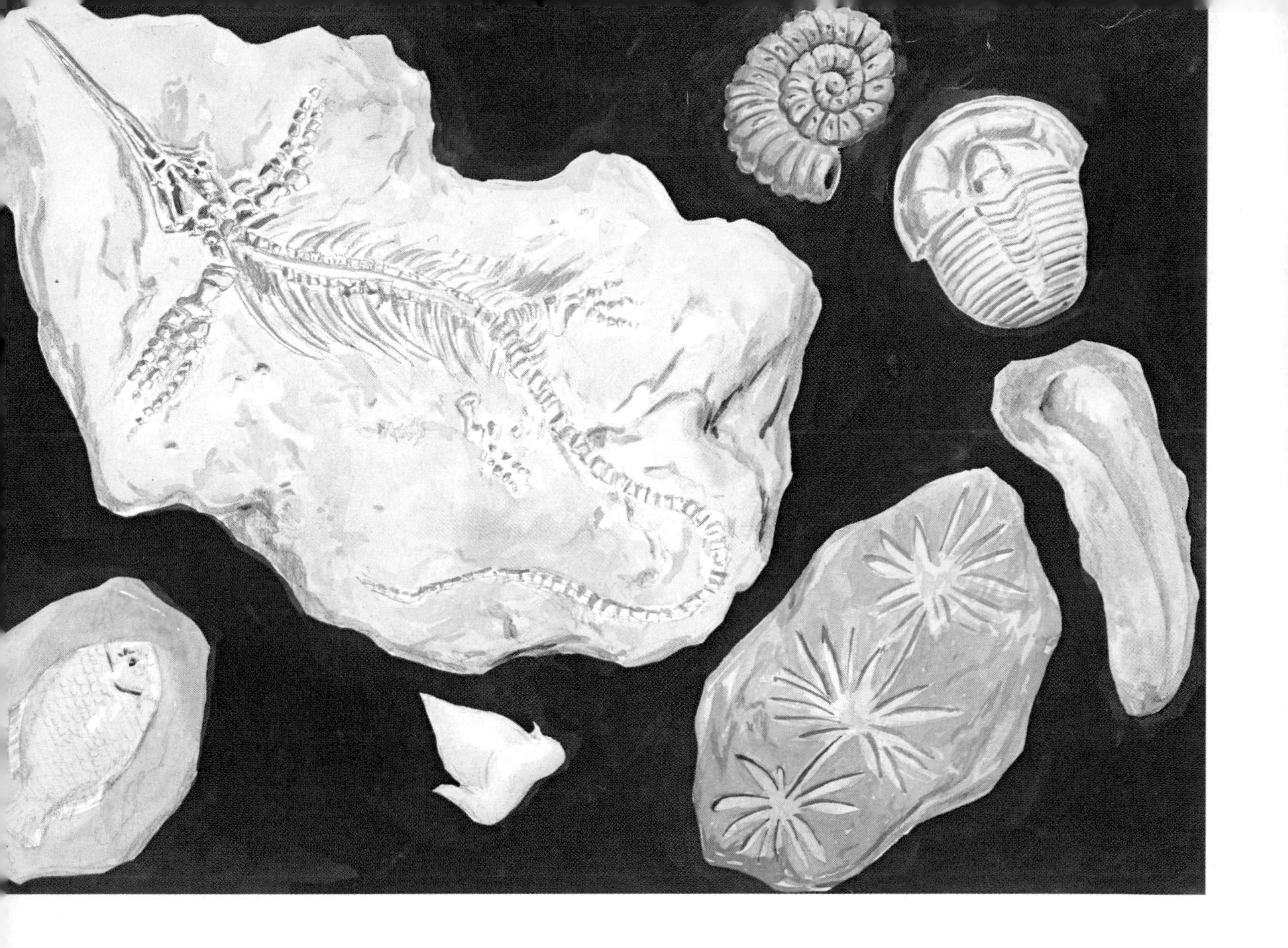

Scientists have learned about pterosaurs by studying the fossil remains of these flying reptiles. Fossils are bones, teeth, or impressions of things that lived millions of years ago. Pterosaurs had such fragile bones that they were often destroyed before they could become fossilized, or turned to stone. So fossils of the flying dragons are rare.

The less a pterosaur weighed, the longer it could glide through the air. This is why the flying reptiles developed light, hollow bones. Although they were hollow, the bones were very strong — especially those of the long "finger" that supported the pterosaur's wing.

In the beginning, the pterosaurs were only about the size of sparrows. With their long spiked jaws, bat-like wings, and long bony tails, they were odd-looking creatures. In the air, they used their slender hind legs to stretch out the skin of their wings. But on the ground, their legs must have been weak and almost useless.

The smallest pterosaurs ate insects — bugs, beetles, and tiny creatures that hopped or flew. Larger pterosaurs speared fish in the lakes and seas. Some probably ate small furry animals, as well as other reptiles. In turn, many pterosaurs were eaten by hungry dinosaurs. Practically helpless on land, they were easily caught by quick-moving meat-eaters like Allosaurus.

Like most other reptiles, pterosaurs hatched from eggs. But pterosaur eggs were very small. How could a baby with long bony wings fit inside a tiny egg? Scientists think the eggs hatched before the babies were completely formed. Until their wings grew larger, young pterosaurs must have depended on their parents for food.

How did a parent bring food to its young? Some pterosaurs had a pouch of skin — something like a kangaroo's. When fish or insects were caught, they could be stuffed into the pouch and carried home to the hungry babies. Other pterosaurs probably carried food in their mouths.

Scientists divide pterosaurs into two groups. In the first group are the early pterosaurs. They all had long tails. One of these was Dimorphodon. With a wingspan of 4 feet, it was about the size of a large hawk. It had a huge skull, and a mouth full of sharp teeth. Soaring on air currents, it searched the skies for flying insects—or even for smaller pterosaurs.

One of the most common of the early pterosaurs was Rhamphorhynchus. Its name means "prow-beak" because it had a beak shaped like the prow, or front, of a ship. Rhamphorhynchus had a wingspread of nearly 5 feet. It also had a very long tail—almost twice the length of its body. The tail ended in a paddle, which was used for steering.

Rhamphorhynchus had very large eyes, and keen eyesight. When it spotted a school of fish in the water below its rocky perch, it took flight. Gliding lower and lower, it skimmed over the water and speared its prey. Once caught, even a slippery fish could not escape from those sharp, forward-slanting teeth.

The second group of pterosaurs were called pterodactyls. There were several different kinds, and they all had short tails, or no tails at all. Some were no larger than sparrows. Others had bodies that were 15 feet long, with huge wingspreads. They had large heads, and long noses that looked like birds' beaks with teeth.

Ctenochasma was a pterodactyl with very long jaws. Fossil skulls show that its long jaws were filled with dozens of teeth. They were almost like bristles. Scientists are not sure of how they were used. One guess is that this ancient reptile may have glided low over the ocean, skimming the water with its lower jaw. The comb-like teeth may have strained tiny creatures from the prehistoric sea.

Over millions of years, the flying reptiles became giants. But their bones and bodies remained light in weight. The big pterodactyls were even better fliers than their ancestors. But one curious thing happened. They lost their teeth! Scientists think this happened to make them lighter while in flight.

The largest of the pterodactyls was the Pteranodon. The name Pteranodon means "winged, but without teeth." These creatures were truly "flying dragons." Some had wingspreads of 30 feet. One huge fossil skeleton measured 51 feet from wing tip to wing tip.

Compared with Pteranodon's large wings, its body was quite small—about the size of a large turkey. Its head was pointed at both ends. In the front was a long, pointed beak. At the back was an odd-looking crest of bone. The crest helped to balance the heavy beak. It may have been used as a rudder while Pteranodon was in flight.

With its light body and large wings, Pteranodon was an excellent glider. Between flights, it rested on cliffs overlooking the sea. Like many other pterosaurs, it probably hung head-down by its hind legs — like a bat. To fly again, it simply dropped from the edge of the cliff into the rising air currents.

Pteranodon probably spent much of its time searching for food. When it saw a school of fish, it flew just above the waves, and grabbed anything close to the surface. This skilled flier expertly rode rising air currents on its long glides. Scientists think Pteranodon's flight control was so perfect that it almost never had to make a forced landing.

But there was danger, even for the best flier. Large reptiles cruised the seas. A hungry Tylosaurus, lurking near the ocean's surface, could make a swift lunge at the low-flying giant. Powerful jaws and jagged teeth could easily rip through the leathery skin and snap the hollow bones. Pulled down into the sea, there was no more hope for a captured Pteranodon.

Along the edges of cliffs, rising air currents made it easy for flying reptiles to launch themselves into the air. But what if they were forced down where there were no cliffs? Scientists say that if there was a tree nearby, a pterosaur may have hopped awkwardly over to it. Then, using its wing-claws, it may have climbed up and launched itself into the air again.

If there were no trees around, a grounded pterosaur might have been able to claw its way to the top of a boulder—or drag itself to the top of a sand dune. Then, wings spread, it might have waited for a gust of wind to take it into the air. Some pterosaurs may have had enough spring in their legs to hop into an updraft of wind that would lift them into the air again.

For millions of years, the pterosaurs—large and small—were kings of the air. Then, about 70 million years ago, all the flying reptiles died out. The great land reptiles—the dinosaurs—also died out. And the monstrous sea reptiles disappeared from the oceans. Scientists think that the world's climate may have changed very quickly, and the reptiles could not survive.

When the colder climates came, the birds managed to survive. Birds were warm-blooded. Birds had feathers to keep them warm. Flying reptiles had no feathers. The colder climate also brought strong winds. Thousands of pterosaurs were probably smashed against cliffs, blown down into the sea, or forced down on land—where meat-eating animals killed them.

Birds had other advantages over the pterosaurs. The pterosaur's wings were attached to its legs, which made walking difficult. So it was easy prey for its enemies when it was on the ground. A pterosaur with a torn wing could not fly, and could starve to death, or be eaten by another animal. But if a bird lost some feathers from its wing, it could continue to fly while it grew new feathers.

And so, the flying dragons that ruled the skies for millions of years, were replaced by the birds. Birds could adapt, but the pterosaurs could not. Scientists now know more about pterosaurs than ever before. But they are still curious about the strange flying creatures that haunted the seas and cliffs of ancient times.